Introduction and Important Information

This book is designed to be used by BLS providers including First Responders, EMT-B, and EMT-IV. We've included additional IV and IM medications at the end of the reference section for AEMT level providers (however, AEMT protocols vary greatly, so we have only covered the most commonly allowed medications). This manual is intended for educational and reference purposes only and is not designed to be a definitive dosage guide. Third-party material should never override local protocols. BLS scope of practice can vary from region to region including rules regarding medication administration. Only act in accordance with your Medical Direction, local standing orders, and regional guidelines.

We chose medications for this book based on the protocols of several BLS manuals provided by EMS organizations in different regions of the United States as well as NREMT study materials for EMT-B and AEMT level providers. Information about medications in the following pages combines FDA labeling data as well as commonly used dosage information mentioned in NREMT and regional standards.

References for all sources can be found at the end of this book. This text was peer reviewed by qualified professionals who can be located by name and title after the reference section.

Bonus Patient Scenarios: Want to apply your knowledge of the medications in this book to patient scenarios? You can access 6 free patient scenarios, each one involving a different intervention/medication at the BLS/EMT-B level. Read the scenario, think about your actions, and then enter any medication you would give as well as expected outcomes in the Run Report. Not only will this help you practice your newly gained medication knowledge, but it will give you a little bit of charting practice. Remember; if it's not in the chart, it didn't happen!

Here's how to access our free patient scenarios in ChartFlow:

1. Go to ChartFlow.io
2. Click to create a "New Student Account"
3. Enter Access Code: **77c3c26294**

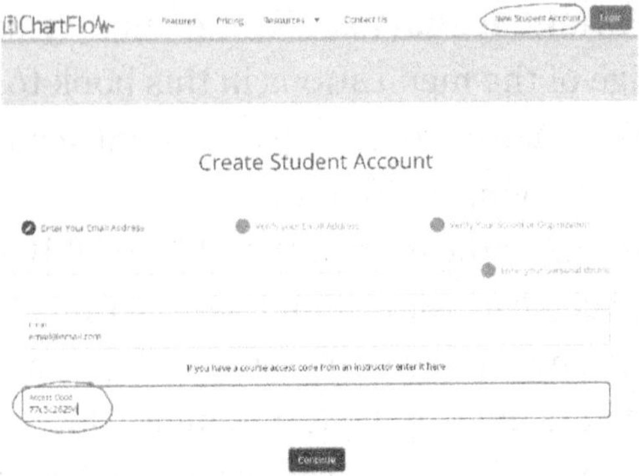

Make sure to enter the Access Code on the first page when you create an Account. If you miss this step, you'll need to contact ChartFlow's support. Once you've created your account, you should see the following screen with activities available.

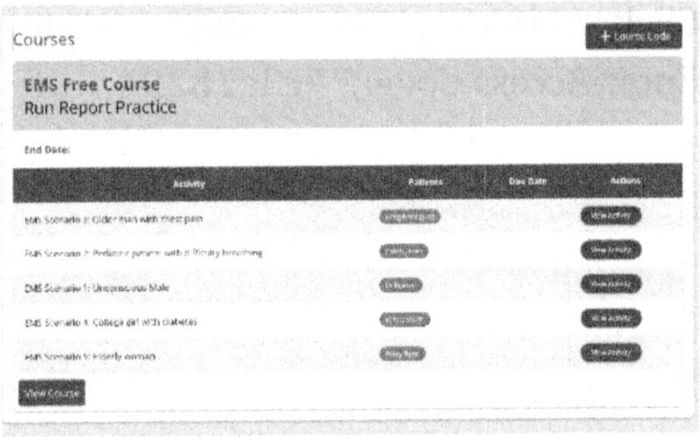

A note on hypersensitivities/allergies:

Hypersensitivities to a medication or associated ingredients are always a contraindication (meaning, you should not give a medication the patient is allergic to). <u>Always</u> ask about allergies before giving any substance.

We chose not to include this warning under every medication as it should be a universal fact when giving substances of any kind. At every level, you will be tested on this as it needs to become reflexive in a real world setting.

A rare, but life threatening concern is the possibility of an allergic reaction (hypersensitivity), which could lead to anaphylaxis.

Signs and Symptoms of Anaphylaxis

- Shortness of breath
- Wheezing
- Stridor
- Chest or throat tightness
- Syncope
- Weak, rapid pulse
- Hypotension

If a patient shows signs of anaphylactic shock, immediately call for ALS support. Use an EpiPen (0.3 mg of epinephrine) if available and administer oxygen via non-rebreather or nasal cannula while in transport to definitive care.

If a patient stops breathing and doesn't not have a pulse, begin CPR.

For adult CPR: Call for advanced life support.

Interlock fingers, lock elbows, and press in the center of the chest.

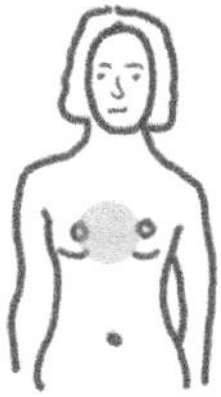

Press hard (at least 2 inches down) and fast (100-120 times a minute) making sure to let the chest completely rise between compressions.

For every 30 compressions do 2 rescue breathes. If rescue breathes are not possible, continue compression only CPR.

Change rescuers every 2 minutes if possible.

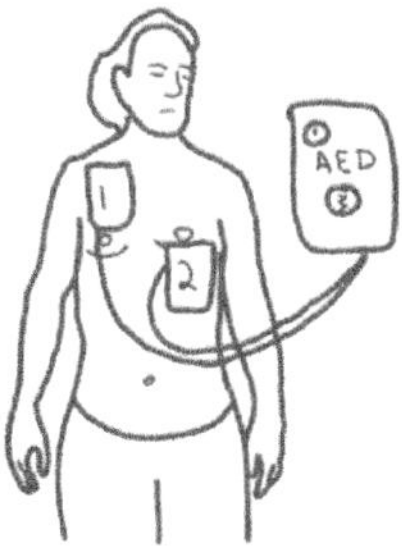

Locate an AED as soon as possible, apply pads and follow machine prompts. If shock is advised, clear for shock and then immediately resume CPR.

Continue the cycle.

Important warning regarding oral medications

Never give an oral medication to an unconscious patient or a patient who cannot protect their airway (evaluate for altered mental status (AMS), swallowing ability, level of consciousness, etc.).

You will see this warning under several medications that often pop up on EMT exams with unconscious or mentally altered patients. We have included it selectively to help students on exams. However, no oral substance should *ever* be administered to a person who does not have the ability to swallow and protect their own airway. This warning should be assumed with all oral medications.

General Medication Administration Guidelines and Tools

Six rights of medication

Always check the six "Rights" of medication before giving any substance.

1. Right Patient - If giving a patient their own medication, verify the patient's name and compare it to the name on their autoinjector or pill container. If you have multiple patients, verify the patient and medical need before using any intervention.
2. Right Medication - Do you have the right medication? Read the label to confirm it matches the intended medication and dose. Check both the outer box and inner vial if applicable. If using a patient's own medication, verify against the original packaging (not a secondary pill container).
3. Right Dose - Confirm that the amount of the pill, liquid or autoinjector matches the dose on the medication label. Have you verified pediatric doses based on weight for

medications like acetaminophen (if allowed by local protocol)? Is the dose in the syringe/pill/liquid the same amount you intended to give?

4. Right Time - Check the expiration date on all medications. For injections or nebulizers, are you giving the medication over the correct amount of time? Have you waited the correct interval of time before giving additional doses? In a transfer setting, does the time you're giving the medication match the orders?
5. Right Route - Verify the route (oral, IM, inhaled, sublingual) before administering any medication.
6. Right Documentation - Always document every medication given. *If it isn't documented, it didn't happen!*

Definitions and abbreviations

Clean technique - Any injections or IV placements should be done after regular hand washing while wearing non-sterile gloves. Always clean the injection site with alcohol or soap.

IM - intramuscular - IM injections are given at a 90 degree angle. IM autoinjectors are used on the outer thigh. Alternate locations may be permitted per local protocol.

IN - intranasal - Intranasal is administered in the nostrils with a pre-filled spray or nasal atomizer.

IO - interosseous - Injected into the bone marrow by inserting a catheter into the bone.

IVP - IV push - Pushing a syringe of medication through an IV line.

PO - by mouth

SC or SQ - subcutaneous - SQ injections are given at a 45 degree angle into the fatty layer beneath the skin.

SL - sublingual - A medication that is placed under the tongue. For example, nitroglycerin.

Quick Weight Conversion Chart

1 lb = .45 kg

1 kg = 2.2 lbs

Pounds	Kilograms
10	4.536 kg
20	9.072 kg
30	13.608 kg
40	18.144 kg
50	22.68 kg
60	27.216 kg
70	31.751 kg
80	36.287 kg
90	40.823 kg
100	45.359 kg
150	68.039 kg
200	90.719 kg

Medications to Know

Acetylsalicylic acid (Aspirin)

Trade Names: Bayer

Drug Category: NSAID

Common Use: Chest pain with cardiac origin, suspected myocardial infarction.

Form: Chewable Tablet, 81 mg

160, 325 mg uncoated tablets are available but less common in EMS

Dose: Up to 325 mg

Administration: Give up to 4 chewable 81 mg tablets. Chewable tablets are usually carried by EMS because they are faster acting than traditional capsules or tablets.

Mechanism of action: As an irreversible COX-1 and COX-2 inhibitor, aspirin binds with and inhibits COX-1 enzymes. These enzymes typically lead to synthesis of pain-signaling prostaglandins. When COX-1 is inhibited, there are less prostaglandins produced. Platelet aggregation is also impaired which can lead to increased bleeding.

Contraindications: GI ulcers, active bleeding, bleeding disorders (ex. hemophilia), aspirin induced asthma, pediatric patients
Side effects: Prolonged and increased bleeding, GI discomfort, Reye's Syndrome, Bronchospasm (especially in asthmatics).
Special notes: Many people will take Aspirin themselves when experiencing cardiac pain. Always ask before giving additional pills to prevent overdose. Use in children is strictly contraindicated because it may cause Reye's Syndrome.
Toxicity: Aspirin toxicity/overdose *is* possible. IV sodium bicarbonate is the antidote and can be given by an advanced level Paramedic or other qualified medical professional.

Acetaminophen (Internationally: Paracetamol)

Trade Names: Tylenol

Drug Category: Analgesic, antipyretic

Common Use: Fever reduction in an alert child post-febrile seizure (not postictal), general fever reduction, pain relief

Form: Oral Suspension, 160 mg/5mL

Dose: 15mg/kg

Weight in lb	Age in years	Dose in mL
Less than 24 lbs	Less than 2 years	Check local protocol
24-35 lbs	2-3 years	5 mL
36-47 lbs	4-5 years	7.5 mL
48-59 lbs	6-8 years	10 mL
60-71 lbs	9-10 years	12.5 mL
72-95 lbs	11 years	15 mL

Administration: Draw up the correct amount in an oral syringe and give to an alert, conscious child who can protect his/her own airway.
Mechanism of action: Likely a COX inhibitor, exact mechanism unknown
Contraindications: Liver disease/failure
Side effects: Nausea, stomach pain, jaundice
Special notes: We have included this as a pediatric medication since it's commonly used for pediatric fevers. Adults will usually take acetaminophen independently for headaches, pain, or fever without the need for EMS intervention.
Toxicity: Acetaminophen overdose/toxicity *is* possible. The antidote is Acetylcysteine/N-acetylcysteine which can be administered at any medical center or hospital.

Activated Charcoal

Trade Names: Liqui-Char, Actidose, Humco

Drug Category: Antidote

Common Use: Poison ingestion

Form: Powder

Liquid Suspension, 25 gram/120 mL and 50 grams/240 mL

Dose: 1 g/kg

For adults and pediatric use.

Administration: Mix powder with water or non-acidic fruit juices.

Mechanism of action: Activated charcoal binds toxins in the GI tract that have not yet been absorbed. The activated charcoal/toxin complex is then eliminated from the body.

Contraindications: Unconscious patients, GI obstruction, corrosives, ingestion of turpentine or petroleum products including gasoline, kerosine, paint thinner.

Side effects: Dark stools, nausea, vomiting

Special notes: If toxins have already been absorbed by GI cells, activated charcoal will not be able to bind with them which is why it should be administered within 2 hours of poison ingestion. Some studies show efficacy for 4 hours if a large quantity of poison was consumed or medications in the form of extended release capsules were taken. Dairy products can decrease effectiveness.

Never give an oral medication to an unconscious patient or a patient who cannot protect their airway (evaluate for AMS, swallowing ability, level of consciousness, etc.).

Albuterol/Salbutamol

Trade Names: Combivent, Proventil, Ventolin, Proair

Drug Category: Bronchodilators, Beta 2 agonist

Common Use: Asthma, COPD with wheezing, emphysema with wheezing, bronchospasm

Form: Liquid 2.5 mg/3 mL normal saline, or metered dose inhaler (MDI)

Dose: If a patient is using their own MDI, the dose is preset.

Adult nebulizer: 2.5 - 5 mg given with a handheld nebulizer over 5-15 minutes.

Pediatric nebulizer: Dosage varies by region. Some examples are 0.15 mg/kg in 2.5-3.0 ml normal saline or 2.5 mg in 3 mL normal saline given with a nebulizer over 5-15 minutes, repeated up to two times.

Administration:

How to set up a nebulizer

1. Open medication vial or single dose pouch and pour it in the nebulizer reservoir
2. Screw on reservoir lid
3. Attach a T piece if applicable

4. On one side of the T piece attach the ribbed hose, on the opposite side attach the mouthpiece
5. Attach tube going to your oxygen tank at the bottom of the nebulizer
6. Set oxygen flow rate to 6 L/min to deliver the dose over 5-15 minutes
7. Have patient put mouthpiece into mouth and breathe normally

How to coach a patient who is using their own MDI

1. Shake well and remove cap. Attach spacer device if available.
2. Breathe out completely
3. Place the mouthpiece between teeth and seal lips around it
4. Press the canister down one time and inhale slowly and deeply
5. Hold breath for 10 seconds before breathing out

Mechanism of action: As a Beta 2 agonist, albuterol stimulates Beta 2 receptors in the sympathetic nervous system and leads to bronchial smooth muscle relaxation and dilation, making it easier to breathe.

Contraindications: Use with caution with extreme tachycardia or MAOI use.
Side effects: Coughing, increased sputum or nasal congestion, dry mouth, tachycardia, palpitations. Paradoxical bronchospasm is possible.
Special notes: In many regions, BLS providers can only assist with a patient's own metered dose inhaler (MDI). However, you may need to coach the patient on how to use their device properly.

Atropine Sulfate and Pralidoxime Chloride

Trade Names: DuoDote

Drug Category: Antidote

Common Use: Organophosphate poisoning from Sarin,Soman, Tabun and VX. These can be used in chemical warfare or as pesticides.

Form: Autoinjector, 2.1 mg/0.7 mL of atropine and 600 mg/2 mL of pralidoxime chloride

Dose: 1 injection IM for people weighing more than 90 lbs or 41kg

Administration:

1. Hold auto-injector firmly in your dominant hand
2. Remove gray cap from top of device
3. Press the auto-injector hard against the recipient's outer thigh until you feel it engage. Hold in place for 10 seconds before removing.

If working in a high risk area, military and health personnel may be instructed to use 1 auto-injector for mild symptoms, and a 2nd auto-injector if symptoms persist. Two additional auto-injectors may be used after an initial single injection if symptoms become severe including altered behavior, confusion, self defecation, convulsions, severe muscle twitching. Multiple dosing may not be included in regional BLS training.

Autoinjector example

Mechanism of action: In organophosphate poisoning, Acetylcholine (ACh) builds up in the body and acetylcholinesterase, which normally breaks down ACh, is inactivated. This leads to excess stimulation of muscarinic and nicotinic receptors. Atropine competitively blocks ACh at muscarinic cholinergic receptors and peripheral autonomic ganglia stopping ACh from acting on receptors around the body. Then, Pralidoxime, reactivates acetylcholinesterase so the excess ACh floating around can be broken down before causing more harm.

Contraindications: If known organophosphate poisoning has occurred, no contraindications should prevent administration in an adult or person over 90 lbs.

Side effects: Dry eyes, urinary retention, dizziness, drowsiness, headache, muscle weakness, dry mouth.

Special notes: Common mnemonics to remember signs of organophosphate poisoning include:

SLUDGE (salivation, lacrimation, urination, diarrhea, GI upset, emesis).

DUMBBELLS (diarrhea, urination, miosis, bronchorrhea, bradycardia emesis, lacrimation, lethargy, salivation).

Mnemonics simplified: Too much ACh = too much stimulation = too much fluid coming out of a body!

Diphenhydramine

Trade Names: Benadryl
Drug Category: Antihistamine
Common Use: Allergies
Form: Oral Solution 12.5 mg / 5 mL, Chewable tablets 12.5 mg
Dose: Adults - 25 - 50 mg by mouth
Pediatric - 1mg/kg up to 25 mg total
Administration: Diphenhydramine is sold as an over the counter medication. If local protocols allow, it is usually given as an oral solution to pediatric patients. To give the oral solution, draw up the appropriate amount in an oral syringe and administer.
Adults and older children can consume the correct number of chewable tablets.
Mechanism of action: Allergens/antigens normally bind to H1 receptors which in turn activates cells in the peripheral nervous system to cause constriction of smooth muscles, bronchoconstriction and dilation of blood vessels. Diphenhydramine is an H1 antagonist and binds to H1 receptors so that allergens cannot. This binding also triggers opposite

effects which decrease the signs and symptoms of an allergic response.

Contraindications: Asthma

Side effects: Drowsiness, sedation, dry mouth, seizure

Special notes: If a patient is having difficulty breathing or other signs of anaphylaxis, epinephrine should be used.

Never give an oral medication to an unconscious patient or a patient who cannot protect their airway (evaluate for AMS, swallowing ability, level of consciousness, etc.).*

Epinephrine

Trade Names: EpiPen

Drug Category: Sympathomimetic agents

Common Use: Anaphylaxis

Form: Autoinjector

Dose: Pediatric - 0.15 mg, Adult - 0.3 mg

Administration:

1. Remove safety cap from autoinjector
2. Press the injection-side against the outer thigh at a 90 degree angle until it clicks
3. Hold for 10 seconds
4. If second autoinjector is available, repeat after 15 minutes if needed

Mechanism of action: Epinephrine is an α and β receptor agonist as well as a histamine antagonist. During anaphylaxis, α agonist action decreases vasodilation so vessels are not as leaky. β1 activity leads to a positive inotropic and chronotropic response. β2 receptor agonist activity relaxes smooth muscles in bronchi, which helps open the airways.

Contraindications: Tachycardia, hypertension, coronary insufficiency, hypothermia, narrow-angle glaucoma

Side effects: Heart palpitations, chest pain, headache, anxiety, hypertension, tachycardia, nausea, restlessness, syncope in asthmatics

Memory Help - Epinephrine is one of the main neurotransmitters of the sympathetic nervous system. Most side effects will mimic what happens during the "fight or flight" response.

Special notes: Beta blockers will reduce the effectiveness of Epinephrine.

Glucose/Dextrose

Trade Names: Glutose, Glucose 15, Insta-Glucose

Drug Category: Carbohydrate

Common Use: Hypoglycemia

Form: Oral Paste

Dose: 12-25 grams for adults, 5-25 grams for pediatrics.

Administration: Squeeze tube directly into mouth or onto tongue depressor to spread against cheek or tongue. Repeat every 5-10 minutes if additional blood glucose readings still show low glucose levels.

Mechanism of action: Increases blood glucose levels by direct consumption

Contraindications: Unconsciousness

Special notes:

Never give an oral medication to an unconscious patient or a patient who cannot protect their airway (evaluate for AMS, swallowing ability, level of consciousness, etc.).

Ibuprofen

Trade Names: Advil, Motrin

Drug Category: NSAID

Common Use: Fever reduction in an alert child post-febrile seizure (not postictal), general fever reduction, pain relief

Form: Liquid suspension 100 mg/5 mL

Dose: Pediatric: 10 mg/kg

Age (yr)	Dose mL
2-3 years	5 mL
4-5 years	7.5 mL
6-8 years	10 mL
9-10 years	12.5 mL
11 years	15 mL

**Taken from FDA website https://nctr-crs.fda.gov/fdalabel/services/spl/set-ids/a3288f7c-1fb3-4013-8f84-f874c4ffb49f/spl-doc?hl=ibuprofen*

Administration: Draw up the correct amount in an oral syringe and give to an alert, conscious child who can protect his/her own airway.

Mechanism of action: Ibuprofen is a COX inhibitor. COX enzymes typically convert arachidonic acid into thromboxanes, prostaglandins, and prostacyclins in the body which cause inflammation. Ibuprofen reversibly inhibits COX enzymes so these chemicals can not be synthesized leading to less inflammation and signaling responses.

Contraindications: Preterm infants, active bleeding, GI ulcers, thrombocytopenia, use of blood thinners

Side effects: Nausea, constipation, diarrhea, headache

Special notes: We have included this as a pediatric medication since it's commonly used for pediatric fevers. Adults will usually take Ibuprofen independently for headaches, pain, menstrual cramps, or fever without the need for EMS intervention.

Ipratropium bromide

Trade Names: Atrovant, Combivent

Drug Category: Bronchodilator

Common Use: Maintenance of asthma, bronchospasm, emphysema, and chronic bronchitis

Form: Metered Dose Inhaler

Dose: 1-2 puffs from metered dose inhaler

Administration:

How to coach a patient who is using their own MDI

1. Shake well and remove cap. Attach spacer device if available
2. Breathe out completely
3. Place the mouthpiece between teeth and seal lips around it
4. Press the canister down one time while inhaling slowly and deeply
5. Hold breath for 10 seconds before breathing out

Mechanism of action: As an acetylcholine antagonist, this medicine blocks cholinergic receptors so they cannot be stimulated by ACh. Normally ACh stimulates smooth muscle cells to create cGMP which then causes contraction of the muscles (which can lead to difficulty breathing). By blocking receptors from ACh stimulation, cGMP decreases and the contraction of smooth muscles decreases, allowing bronchioles to relax.

Contraindications: Hypersensitivity to soybeans, narrow angle glaucoma, urinary retention

Side effects: headache, dizziness, nausea, paradoxical bronchospasm, dyspnea

Special notes: This medication is designed as a maintenance medication and is not intended as a rescue medication.

In many regions, BLS providers can only assist with a patient's own metered dose inhaler (MDI). However, you may still need to coach a patient on how to use their device properly.

Naloxone hydrochloride

Trade Names: Narcan, Suboxone, Targin

Drug Category: Antidote

Common Use: Opioid overdose with respiratory depression

Form: Solution for injection or nasal spray

Dose:

Intranasal - 1 mg/nare

Injection - Prefilled vial or autoinjector 0.04 mg with repeated dosing every 5 minutes as needed, up to 2 mg.

Pediatric - 0.1 mg/kg

Administration: Some Naloxone kits include a nasal atomizer. This may come pre-assembled or you may need to replace the tip of a pre-filled syringe with the atomizer.

How to assemble an atomizer attachment:

1. Draw up 2 mg of Naloxone hydrochloride into a syringe.
2. Remove the syringe needle and attach the nasal atomizer, which looks like a cone shaped sponge.

3. Insert the atomizer attachment into the patient's nostril and squeeze 1 mg out of the syringe. Repeat on the second nostril.

Narcan brand naloxone may come in a pre-packaged nasal spray. See blow:

If using a pre-packaged nasal spray, remove it from the wrapper immediately before use.
Wear gloves.
Put middle and index finger on sides of sprayer.

Lie patient down.
Insert nozzle into nostril.
Depress plunger with thumb.
Do not prime medication before use

For IV/IO administration - follow local protocol.

Mechanism of action: As a competitive inhibitor, Naloxone binds to opioid receptors throughout the body. This reverses the effects of opioids since they can no longer bind to receptors. Because opioids are no longer acting on receptors, a long time opiate user may suddenly be launched into withdrawal. However, if a person has not used an opiate drug, they will not have any significant side effects from Naloxone.

Contraindications: In the case of an opioid overdose with respiratory depression there are no contraindications.

Side effects: Combativeness, vomiting, tachycardia, seizures, headache, altered mental status.

Special notes: In opioid dependent people, Naloxone can cause instant withdrawal. Be aware that a patient may become conscious and be extremely combative because they are now experiencing full withdrawal. Do not use this drug on patients who are conscious, able to maintain their own airway, breathe at a normal rate and rhythm, and have no other life threatening problems.

Nitroglycerin

Trade Names: Nitrostat, Nitro-Bid, Nitrolingual
Drug Category: Vasodilator
Common Use: Chest pain with suspected cardiac cause, myocardial infarction, angina pectoris, congestive heart failure with pulmonary edema
Form: Oral tablet 0.4 mg, Sublingual spray 0.4 mg/spray, topical ointment 7.5 mg/half inch
Dose: 0.4 mg
Administration:
Sublingual tablets and spray -

1. Give one .4 mg sublingual tablet or spray by placing under the tongue.
2. Wait 5 minutes.
3. If pain persists and systolic heart rate is still above 100 beats per minute (please note, some protocols use 90 beats per minute as the safety marker), administer another dose.
4. Wait 3-5 more minutes.
5. Repeat once more for a total of 3 sublingual sprays or tablets in total.

Topical ointment - Wear gloves to prevent exposure to medication. Measure ½ inch of medication on paper applicator and spread it on a hairless portion of the patient's skin.

Topical application is typically used for daily angina maintenance. Sublingual administration is preferred in EMT response when the patient has tablets or sprays available.

Mechanism of action: Nitroglycerin is converted to Nitric Oxide (NO) in smooth muscles which leads to vasodilation. Arterial dilation causes decreased afterload and venous dilation causes decreased preload.

Contraindications: Hypotension, recent use of phosphodiesterase 5 (PDE5) inhibitors in the last 24-48 hours (Viagra, Cialis, Levitra)* hypovolemia, bradycardia, increased intracranial pressure

Side effects: Headache, decreased blood pressure, reflex tachycardia, syncope, nausea, vomiting

Special notes: If a person takes a PDE5 inhibitor before taking nitroglycerin, they can experience a dangerous drop in blood pressure. Be sure to ask all patients about PDE5 inhibitors regardless of gender since they can be used for more than erectile dysfunction.

**Some protocols require 24 hour since last use of Viagra, and 48 hours since the last use of Cialis and Levitra*

Ondansetron

Trade Names: Zofran

Drug Category: Antiemetic

Common Use: Nausea and vomiting

Form: Quick dissolving tablet 4 mg

Dose: For patients over 40 kg: 4 mg

Administration: Place the tablet on the patient's tongue and advise them to swallow as the tablet dissolves (do not swallow the tablet whole). A second dose may be given after 15 minutes if no improvement.

Mechanism of action: As a selective 5-HT 3 (serotonin) receptor antagonist, Zofran blocks serotonin from acting on receptors. This is thought to prevent nausea that was caused by increased serotonin, but the mechanism of action is not fully understood.

Contraindications: Phenylketonuric (PKU), long QT syndrome, use of apomorphine for Parkinson's

Side effects: Headache, drowsiness, weakness, dry mouth, constipation, diarrhea

Gasses

Oxygen

Common Use: Hypoxia, shock, respiratory distress, cardiac distress, carbon monoxide poisoning, anemia, shortness of breath, low SPO2

Dose and administration: Low flow = 1-2 L/min. Medium flow = 4-6 L/min High flow = 10-15 L/min

Administration: Flow guidelines below:

Nebulizer	6-8 L/min
Nasal Cannula	1-6 L/min
Non rebreather mask (with respiratory distress)	10-15 L/min
Bag Valve mask (breathing insufficient or absent)*	15 - 25 L/min
CPAP (COPD, CHF with pulmonary edema, asthma)	5-15 cmH20

**Normal respiratory rate for adults is 12-20 per minute.*

Mechanism of action: Oxygen attaches to heme groups in hemoglobin and is carried by red blood cells to cells throughout the body.
Side effects: Light-headedness, dry nose or throat
Special notes: Patients with long term COPD may become hypocapnic with high flow oxygen. Observe all patients for changes in respiratory rate and be prepared to assist with a bag valve mask as needed. In a CVA or head injury, SPO2 should not be increased above 94%.
Oxygen tanks are highly flammable. Handle with care.

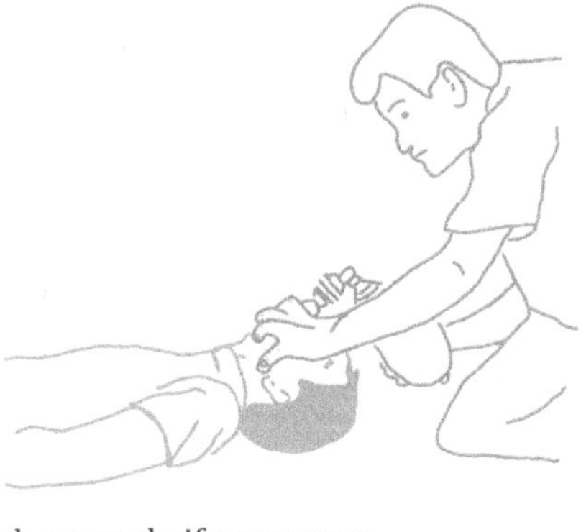

Bag valve mask example if necessary

Nitrous Oxide (AEMT level)

Common Use: Pain from severe trauma

Dose: Delivered in 50/50 mix with oxygen. Typically self administered until pain is relieved.

Nitrous Oxide must be given with oxygen as it can cause mild respiratory depression leading to hypoxemia

Administration: Use a tightly sealed mask or a face shield attachment.

Mechanism of action: Nitrous Oxide may inhibit NMDA receptors which creates an anesthesia effect, and activates GABA-A receptors for an anxiolytic effect.

Contraindications: COPD, pneumothorax, bowel obstruction, decompression sickness, intraocular gas, middle ear gas

Side Effects: Nausea, vomiting, dizziness, weakness, confusion, hallucinations, laughing

Special Notes: Nitrous Oxide diffuses quickly into free air cavities (for example, the bowels) and can cause quick expansion of the air in any given space. It's important not to give Nitrous Oxide to a person with a pneumothorax.

IVs and Injections

IVs are typically reserved for EMT-IV and AEMT level providers. Before listing IV medications, we have included some definitions and tips related to IV and IO insertion.

Slow Intravenous Push (Slow IVP, SIVP) - "Slow" IV push can be defined as an IV push over 1-2 or 3-5 minutes depending on the provider. We use the term "slow IV push" in this manual as an IV push done over multiple minutes, the exact time to be determined by local medical direction.

TKO (to keep open) - TKO rate also varies by region with 25 ml/hr being a commonly used rate

Flow rate calculation

$$\frac{\text{volume in mL}}{\text{time in minutes}} \times \text{drip set} = \text{flow rate}$$

Typical gauge size for different IV catheter insertions

Adult patient	18 - 20 gauge
Adult trauma patient	14 - 18 gauge
Pediatric	20 - 24 gauge

Inserting an IV catheter and lock

This is a skill every EMS provider with IV rights should practice during training. Some important points to review are listed below:

- Wipe the insertion area with alcohol before inserting a catheter needle
- Use proper disposal for sharps to avoid needle stick injuries
- Attach a lock only after flushing it
    - Attach the cap of the "lock" to a 10 mL normal saline syringe
    - Depress the syringe slowly until fluid drips from the lock tube (still unattached to the IV catheter)
    - Attach the lock to the catheter

- Once a lock is in place, additional bags of normal saline or lactated ringers may be connected to the correct tube on the lock

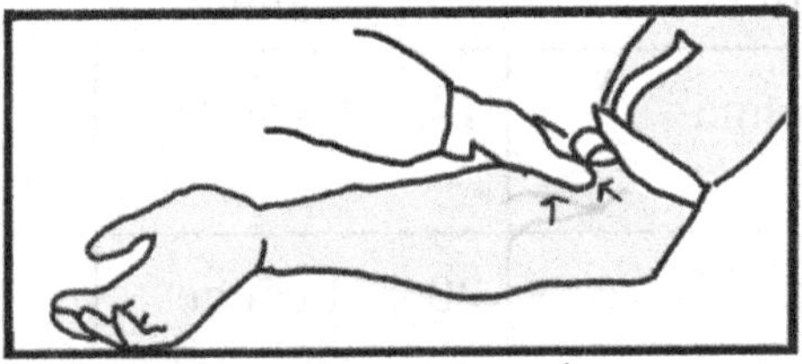

Pull the skin taught after cleaning site

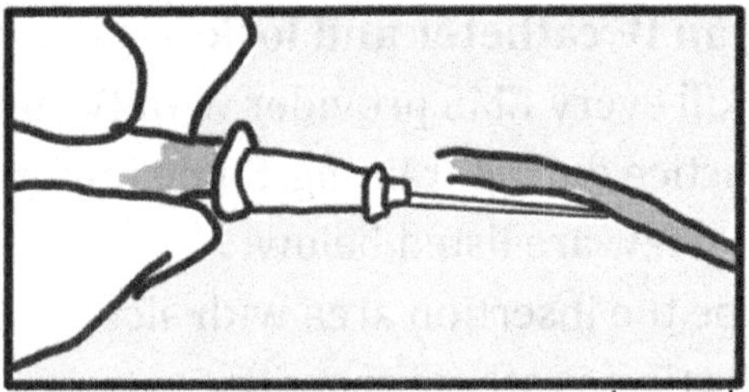

Insert needle at 30 degree angle until blood enters

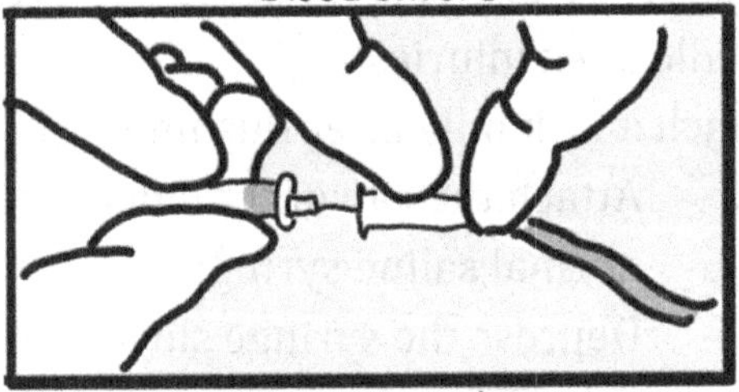

Remove needle
Advance and secure catheter

Inserting IO needle/catheter

IO access can be obtained quickly using the following steps. This can be a painful procedure and IV lidocaine can be used if local protocol allows.

1. IO is typically done in the Proximal Tibia (preferred for pediatrics) or Proximal Humerus. Secure limb for minimal movement.
2. Apply aseptic technique including sterile gloves, wiping the insertion site with an antiseptic, and using sterile supplies.
3. Place the IO needle about 2 cm below the tibial tuberosity on the flat anteromedial surface of the proximal tibia. Follow manufacturer instructions for IO needle and catheter insertion.
4. Once you've pushed the needle through the bone to the point you no longer feel resistance, attempt to aspirate a small amount of blood. If you can not aspirate anything, insert a 10 mL normal saline flush and observe for signs of SQ swelling or fluid leakage. If SQ swelling or leakage is present,

remove the needle and attempt at a different site.

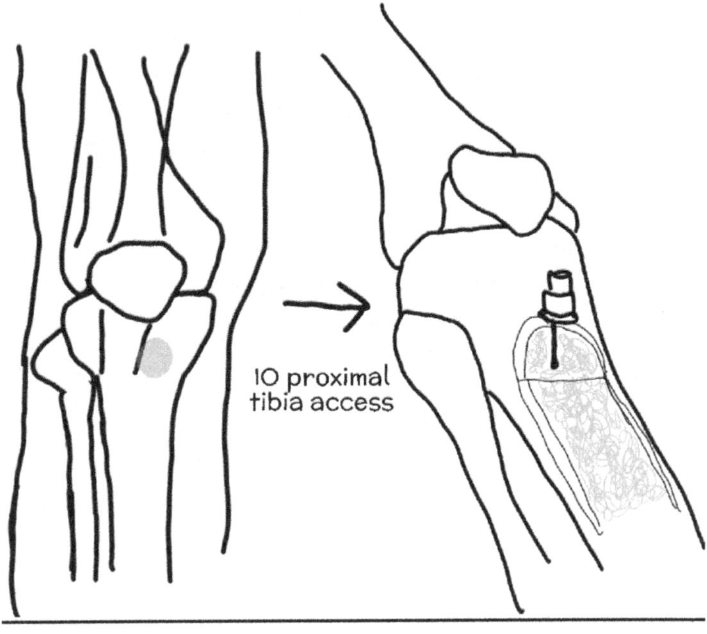

Normal Saline (.9% NaCL)

Drug Category: Isotonic Crystalloid Solution

Common Use: Dehydration, heat exhaustion, heat stroke, hypovolemia, diabetic ketoacidosis, flushing IV locks

Form: Bag with liquid 250 mL, 500 mL, 1000 mL, Prefilled syringe 10 mL

Dose and administration: Attach drip set to bag of Normal Saline (NS). Hang bag or hold above the height of the patient. Control the flow rate as needed.

Mechanism of action: Isotonic solution increases the volume of plasma by directly infusing fluids into the circulation

Contraindications: Severe CFS, renal failure with dialysis

Side effects: Fluid overload possible

Lactated Ringers

Drug Category: Isotonic Crystalloid Solution
Common Use: Burns, hypovolemic shock
Form: Bag with liquid 250 mL, 500 mL
Dose and administration:
Adult: 20 mL/kg
Pediatric: 10 mL/kg
Mechanism of action: Isotonic solution including NaCL, sodium lactate, potassium, and calcium increases the volume of plasma and electrolytes by directly infusing fluids into the circulation.
Contraindications: Severe CHF, renal failure
Side effects: Fluid overload possible

Dextrose

Trade Names: D50, D25, D10

We have included all Dextrose %s in one entry (50%, 25%, 10%). Some services carry only one type and you may have to dilute a vial with sterile saline. For example, dextrose at 50% (D50) dilutes 1:1 to make 25%

Drug Category: Carbohydrate

Common Use: Symptomatic or severe hypoglycemia

Form: 25 grams/50 mL D50 vial or prefilled syringe
50 grams/500 mL D10W bag

Dose and administration:

Adult 25 grams at 50%, slow IV push

Pediatric: 0.5-1g/kg of 25% (dilute with sterile saline 1:1), slow IV push

Infant: Varies by region. Common dose is 2 mL/kg (0.2 g/kg) of 10% Dextrose, slow IV push

**IO push can be used if IV access cannot be established*

Mechanism of action: Dextrose is essentially the same as glucose. Because you are injecting it into the circulation, it quickly increases serum glucose levels.

Contraindications: Intracranial hemorrhage, head trauma

Side effects: Nausea, vomiting, headache, burning at the site of injection, change in heart rate
Special notes: When dextrose leaks into nearby tissues (extravasation) it can cause necrosis. Confirm IV/IO line is clear and properly inserted by flushing the line with normal saline before administration.

<u>Epinephrine (Cardiac)</u>

Drug Category: Sympathomimetic agents

Form: Liquid IV/IO

Dose and administration for cardiac arrest (see autoinjector entry for anaphylaxis treatment):

Adult - 1 mg (1:10,000 or 1mg:10ml) every 3-5 minutes,

Pediatric - 0.01 mg/kg (1:10,000 or 1mg:10ml) every 3-5 minutes

Common Use: Anaphylaxis (see autoinjector entry), cardiac arrest, heart block, severe bradycardia that does not respond to other interventions

Mechanism of action: α and β receptor agonist and histamine antagonist. During anaphylaxis, α agonist action decreases vasodilation and increases vascular permeability. β1 activity leads to positive inotropic and chronotropic response. β2 receptor agonist activity relaxes smooth muscles in bronchi.

Contraindications: tachycardia, hypertension, coronary insufficiency, hypothermia, Narrow-angle glaucoma

Side effects: Heart palpitations, chest pain, headache, anxiety, hypertension, tachycardia, nausea, restlessness, syncope in asthmatics

Memory help - Epinephrine is one of the main neurotransmitters of the sympathetic nervous system. Side effects will mimic what happens during the "fight or flight" response.

Special notes: Beta blockers will reduce the effectiveness of Epinephrine.

Glucagon

Trade Names: GlucaGen

Drug Category: Hypoglycemia antidote

Common Use: Hypoglycemia, overdose of beta blockers or calcium channel blockers

Form: Powder or mixed liquid

Dose and administration:

Hypoglycemia: 0.5 - 1mg IM followed by IV access and 50% dextrose administration

Beta-blocker or calcium channel blocker toxicity: 3 to 10mg IV push over 3 to 5 minutes. Then give supplemental dextrose/glucose.

Mechanism of action: Glucagon is naturally present in the body as an insulin antagonist. It triggers cells in the liver to convert stored glucose (glycogen) into a usable form (glucose) and then releases it into your bloodstream via a process called glycogenolysis.

Contraindications: Glucagon is often used as a last option for hypoglycemic patients who are unconscious. In this situation there are no contraindications.

Side effects: Nausea, vomiting

Special notes: A patient must be able to receive dextrose PO or IV after glucagon administration.

Vaccines

Some regions allow AEMTs and EMT-IV to administer vaccines in some cases. The flu, coronavirus, DTaP, HepA, HepB, Hib, HPV and Tdap vaccines are all given intramuscularly at the time of writing.

Administration:

1. Review vaccine instructions and insert before administration.
2. Always wear gloves.
3. Have patient sit with arm relaxed at side.
4. Locate the deltoid muscle by moving 2-3 finger widths below the acromion process on the upper arm.
5. Wipe the area with alcohol and allow it to air dry. Do not wipe or blow on it.
6. Use the outside edge of your non-dominant hand, from wrist to pinky, to apply moderate pressure and slightly pull the skin away from the injection site.
7. Use your dominant hand to insert the syringe at a 90 degree angle.

8. Depress plunger at a rate of about 10 seconds per mL.
9. Cover injection site with gauze or bandage.

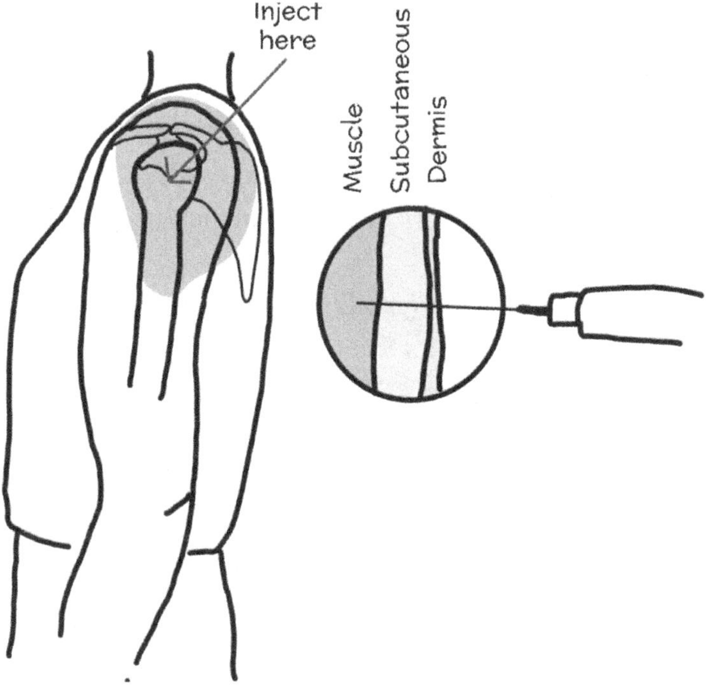

References

The majority of the medication information listed in this guide is checked against the Food and Drug Administration website. References to physical skills come from the writer's education as an EMT-B. Other sources of protocols and guidelines are listed below.

Food and Drug Administration. (n.d.). FDALabel. https://nctr-crs.fda.gov/fdalabel/ui/search

@LightsSirensAction (2022, May 1). Medication Monday [Lights Sirens Action EMS]. YouTube. Retrieved May 16, 2023, from https://www.youtube.com/playlist?list=PLzrNkrNeqKv-q28PDX1boCEEg17UEEBZb

Mesa County. (2019, March 1). Mesa County EMS System Guidelines. Mesa County specific EMS Protocols V 19.1. https://ems.mesacounty.us/mcweb/ems/protocols/19.1/

National Registry of Emergency Medical Technicians. (n.d.). National Continued Competency Program. National Registry of Emergency Medical Technicians. Retrieved May 16, 2023, from https://content.nremt.org/static/documents/2016_NRP_NCCP_final.pdf.

South Dakota Department of Public Safety. (2010). EMT-Basic Pre-Hospital Treatment Guidelines 3 rd Edition. Pierre , SD; South Dakota Department of Public Safety. Retrieved May 16, 2023, from http://dps.sd.gov/emergency_services/emergency_medical_services.

West Virginia Department of Health and Human Resources. (2023). BLS Treatment Protocols. West Virginia Office of Emergency Medical Services. Retrieved May 16, 2023, from https://www.wvoems.org/media/457666/2023%20emt%20protocols%20for%20print.pdf.

Special thanks to the people who helped review this book for accuracy:

Jocelyn Wensel, MA
Jocelyn is a second year medical student at International University of Health Sciences.

Perry Putnam, MPA
Perry Putnam is the Class Coordinator and Lead Instructor for Everett Community College's Emergency Medical Technician (EMT) program. Perry holds a Bachelor of Science degree in Paramedicine, a Bachelor of Arts degree in Law and Justice, and a Master degree in Public Administration. He has also completed the 4-year Executive Fire Officer Program from the National Fire Academy in Emmitsburg, Maryland. When he is not instructing at Everett Community College, he serves as a Battalion Chief Snohomish Regional Fire and Rescue, where he started as a paramedic in 1996.

Made in the USA
Monee, IL
22 October 2026

66441481R00036